HowExpert Presents

Oil & Gas Drilling Guide

HowExpert with Miguel Ferraz

For more tips related to this topic, visit HowExpert.com/oilgas.

Recommended Resources

- HowExpert.com – Quick 'How To' Guides on All Topics from A to Z by Everyday Experts.
- HowExpert.com/free – Free HowExpert Email Newsletter.
- HowExpert.com/books – HowExpert Books
- HowExpert.com/courses – HowExpert Courses
- HowExpert.com/clothing – HowExpert Clothing
- HowExpert.com/membership – HowExpert Membership Site
- HowExpert.com/affiliates – HowExpert Affiliate Program
- HowExpert.com/writers – Write About Your #1 Passion/Knowledge/Expertise & Become a HowExpert Author.
- HowExpert.com/resources – Additional HowExpert Recommended Resources
- YouTube.com/HowExpert – Subscribe to HowExpert YouTube.
- Instagram.com/HowExpert – Follow HowExpert on Instagram.
- Facebook.com/HowExpert – Follow HowExpert on Facebook.

Table of Contents

Chapter 1: Pore Pressure & Fracture Gradient

Pore Pressure Prognosis

The first step to drill a well is to predict the pore pressure and fracture gradient that the well will encounter. In most cases the information that is available is very limited, so the main difficulties with oil and gas drilling start right at the beginning. Owing to this fact, some assumptions have to be considered. In new frontier areas, no information is available, only seismic and sometimes in very bad quality. Therefore, only an educated guess is possible in terms of determining the pore pressure of the formations that will be encountered while drilling. In not such frontier areas, some information can be available like formation tests that provide us with the pressure that was encountered at different depths. The formation tests data can be used to calculate the pore and fracture pressures of the formations above those pressure points, once again assuming certain values.

For example, the pressure data, can indicate that the gradient from the pressure points to mean sea level is a salt water gradient of 0.45 psi/ft, which for example suggests that no abnormal pressures will be encountered. However, it may have a 1,000 psi increment of pressure, below the reservoir top, from 5,000 psi (at 10,000 ft TVD) to 6,000 psi, which could be the known pressure of a reservoir that was already drilled. In this case, we know the pressure of the reservoir but in some other cases we don't. It is important to be aware of all of these uncertainties when we go to drill for oil and gas. In this case, we would assume the reservoir is over-pressured and this needs to be taken into consideration for the pressure predictions. Consequently, it will affect the whole well design as well as the rig selection.

Fracture Gradient Prediction

In terms of the fracture gradient prediction, two models are usually considered. One of them is the Ben Eaton formula, which calculates the fracture pressure from the pore pressure and geostatic pressure as well as the Poisson ratio. Once again, to obtain a predicted value of the fracturing pressure, the following assumptions have to be incorporated:

- Poisson ratio = 0.3;
- Geostatic Pressure needs the following assumptions to be calculated:
  - Formation gradient = 1 psi/ft;
  - Salt water gradient = 0.45 psi/ft.

An alternative, to cross check your information and make sure you are obtaining sound values, is to obtain local data from wells drilled around the area. Obviously, this data needs to be looked at with a critical eye to ensure it is data from wells that have similar formations (rock) to the ones we will be drilling, in other words ensure they qualify as proper analogue wells. For example, a North Sea typical formation fracture gradient is 0.81 psi/ft, and this could be used as a reference if you were to drill a well in the North Sea.

Pressure charts are usually the final outcome of this exercise, with the predictions of the pore pressure, fracture pressure and overburden pressure gradients that, supposedly, will be encountered at specific depths throughout the drilling of the well.

Mud Window

The whole purpose of having pore and fracture pressure curves, is to have a minimum and a maximum in which we can use a certain density of drilling fluid, also known as mud, to drill the well safely, not exceeding the fracture gradient and not going below the pore pressure. This zone between the upper and lower limits is usually called a "mud window". The mud window is given a safety factor to ensure that the limits are not reached while drilling. Usually, the mud window is displayed in a graph with the mud density in the x-axis and the depth of the well in the y-axis, where the pressure and gradient curves are also shown and the mud window is in between those curves.

After obtaining the mud window, we can try to do some standardization of a hypothetical well outline. The following assumptions can be applicable to development wells within a field that already drilled its exploration and appraisal wells, i.e., it is possible to confidently elaborate a well design by applying these assumptions. Note be taken that it is possible to obtain this standardization, when an intensive regional geology research is done with good quality data as well as a significant effort is undertaken to understand local well design architecture.

Considering all this, it would be possible to assume tentative casing setting depths, which will later be analyzed and confirmed, or not, in the casing design process. A possible standard configuration for an offshore project with about 300 ft of water depth, nowadays considered a shallow water depth, without any special anomalies in the formations that will be drilled, could have this tentative well configuration:

- 26” Phase – 20” Conductor Casing: Casing Depth 300 ft below WD;
- 17-1/2” Phase – 13-3/8” Structural (or Surface) Casing: Casing Depth at 2,000 ft, since it was assumed that no drilling hazards were presented in the respective casing setting depth formation;
- 12-1/4” Phase – 9-5/8” Intermediate Casing: Casing Depth 300 ft before top of Reservoir Formation;
- 8-1/2” Phase – 7” Production Liner: Liner Setting Depth should include a 150 ft rat hole (after the end of the reservoir) to allow wireline logging and possible well testing operations.

Note that to drill an oil and gas well, you have to start by drilling with a bigger bit, then you lower a big casing and then you go to the next phase, or also known as hole section, and you drill with a smaller bit and so on lowering a smaller casing and drilling with an even smaller bit until you reach the reservoir, which is your main objective. The reservoir is the section of your well that has a permeable and porous rock, hopefully with oil or gas, which are both made out of hydrocarbons. If hydrocarbons are encountered, they are lying in the pores of the rock and the reservoir rock needs to have enough permeability between its pores for the hydrocarbons to flow into the well and to be produced, which is the ultimate objective of oil and gas drilling.

Chapter 2: Well Configuration

The well configuration is the decision-making process of determining what is the best fit for purpose configuration for the well you want to drill. Considering that you have specific targets to reach, but you need to reach those targets in a safely manner. This whole process can be divided in three consecutive steps. The first step is to define the well trajectory. The well trajectory is defined from knowing two things: i) the exact position of the targets that the geologists want to reach with the corresponding coordinates and, ii) the exact place where the rig will be located at surface. After obtaining these two pieces of information, a well trajectory is defined and optimized. The second step is to understand how we are going to get from surface to the final target in a safely manner. This being said, to reach the final target safely we need a proper notion of the casing setting depths as well as their dimensional requirements. Here is where the casing design process comes in and all worst scenarios are considered to be on the safe side and to ensure the operational crews are never put at risk while drilling the well. The third and final step is to elaborate the cement program, after having the casing program in place, and ensure the cement design is adequate and to gain clarity on the cement that will be necessary to reach the final target in safe conditions.

Well Trajectory

The well trajectory is always a subject of great discussion. The main debates are around the geological targets and their corresponding tolerances in all possible directions, and also how the reservoir engineers want to intersect that target. Therefore, it is a multi-disciplinary task that involves a lot of iterations and prioritizations. It is very easy to have everything on paper but as a drilling engineer, you need to transform what is on paper on something feasible while trying to satisfy all parties. Obviously, some preferences will have to be compromised, and therefore, brainstorming exercises such as SWOT analysis, risk assessments and discussions with specialists are performed to ensure all angles are being considered. It is not an easy task, but it is one that always has to be performed when drilling for oil or gas.

In the oil and gas industry there are quite a few different well trajectories. The most standard ones are: starting with the typical vertical well, and then the J-shape well, the S-shape well and the horizontal well. These are the four main types of wells encountered around the world.

Casing Design/Dimensions

The casing design process, in most cases, reflects the worst case scenario of burst and collapse of the casing landing conditions, as well as, several other load cases. In order to get more details on the calculations, a more detailed Drilling Engineering Manual should be read.

The first step of the casing design exercise is to determine the load cases design factors. In order to calculate the load cases, some design factors have to be considered. The following design factors are suggested by the author (in parenthesis are values suggested by a Drilling Engineering Manual):

- ✓ Burst – 1.1 (between 1.0 and 1.33)
- ✓ Collapse above cement – 1.0 (between 1.0 and 1.125)
- ✓ Collapse below cement – 1.0 (between 1.0 and 1.125)
- ✓ Tension – 1.3 (between 1.0 and 2.0)
- ✓ Triaxial – 1.25 (1.25)

The casing design is calculated for various scenarios, but usually the worst case scenario is for the intermediate casing, which is in most wells the heaviest casing and the casing that is set, just before entering the reservoir. The worst design scenario, which is usually used for the collapse of the intermediate casing, is when the casing is fully evacuated due to lost circulation while drilling. In this case, the casing is empty on the inside and the pore pressure is acting on the outside. On the other hand, the maximum burst load is experienced if the well is closed in after a gas kick has been experienced. The pressure inside the casing is due to formation pore pressure at the bottom of the well, usually being reservoir pressure, and a column of gas which extends from the bottom of the well to surface. It is assumed that pore pressure is acting on the outside of the casing.

Cement Design

After having the well trajectory defined and the casing program finished, the final step of the well configuration is the cement program. Cement plays a big role in the well integrity, in the safety of the operational teams and in avoiding any environmental damage from the well. First, lets understand the functions and principles of the cement in an oil and gas well and then get acquainted with industry best practices including how it should be handled.

Cement is used as a seal between the casing and the borehole. The most important functions of a cement sheath between the casing and the borehole, are:

- To prevent the movement of fluids from one formation to another or from the formations to surface through the annulus between the casing and the borehole.
- To support the casing string (specifically the surface casing).
- To protect the casing from corrosive fluids in the formations.

The prevention of fluid migration is the most important function of the cement sheath between the casing and the borehole. Cement is only required to support the casing in the case of the surface casing, where the axial loads on the casing, due to the weight of the wellhead and BOP connected to the top of the casing string, are extremely high.

Among the variety of classification of cement powders approved for oil and gas drilling applications, by the American Petroleum Institute - API, a Class G cement powder is adequate for most drilling projects. It is a general purpose cement powder compatible with most additives and can be used over a wide range of temperature and pressure.

Pilot tests are usually carried out before the actual cement job is performed, to simulate downhole conditions and ensure the cement behaves according to its properties. The mix-water requirements and additives used play a big role in the cement behavior, and depending on the well conditions and objective of the cement job, cement additives can be used to:

- ✓ Vary the slurry density.
- ✓ Change the compressive strength.
- ✓ Accelerate or retard the setting time.
- ✓ Control filtration and fluid loss.
- ✓ Reduce slurry viscosity.

The main cause for poor isolation after a cement job is the presence of mud channels in the cement sheath, in the annulus. These channels of gelled mud exist because the mud in the annulus has not been displaced by the cement slurry. This can occur for many reasons, the main reason for this is poor centralization of the casing in the borehole, during the cementing operation.

So, in order to improve mud displacement and obtain a good cement bond the following practices are recommended:

- Use centralizers, especially at critical points in the casing string. A typical program for the centralizers is: 1 centralizer immediately above the shoe, 1 every joint on the bottom 3 joints, 1 every joint through the production zone, 1 every 3 joints elsewhere.
- Before doing the cement job, condition the mud to ensure it has good flow properties, so that it can be easily displaced. Wipers/scratchers shall also be run on the outside of the casing to remove mud cake and break up gelled mud. Also a cement spacer fluid shall be pumped ahead of the cement slurry.

The evaluation of the cement job is also paramount. Bear in mind that a cement job shall isolate the undesirable zones. A good cement job occurs if:

- The cement fills the annulus to the required height between the casing and the borehole.
- The cement provides a good seal between the casing and borehole and fluids do not leak through the cement sheath to surface.
- The cement provides a good seal at the casing shoe.

The detection of the Top of Cement (TOC) and the quality of the cement bond can be measured by using cement bond logs (CBL). The CBL tools have become the standard method of evaluating cement jobs since they detect the TOC and also indicate how good the cement bond is. The CBL tool is basically a sonic tool which is run on wireline and emits sound waves that go through the casing, then through the cement and then through the formation and back to the tool. The signal that is received on the tool is then treated and its different responses show the different qualities of the cement bond.

Chapter 3: Drillstring Design

As a starting point, the design of the drillstring is performed in order to meet the following four basic requirements:

- ✓ The burst, collapse and tensile strength of the drillstring components must not be exceeded.
- ✓ The bending stresses within the drillstring must be minimized.
- ✓ The drill collars must be able to provide all of the weight required for drilling.
- ✓ The bottom hole assembly (BHA) must be stabilized to control the direction of the straight sections of the well.

The exercise is easier to understand, if divided into three sections, so let's start from the bottom to the top. First, the bit, then the BHA and last but not least the drill pipe.

Bit Selection

The bit selection is also another complex process, especially in exploration wells, where the formation that will be encountered is unknown. Therefore, as a starting point, it is imperative to get as much information as possible on the formation that is going to be drilled even if it is minute, because when the Request for Quotation (RFQ) is sent out to the vendors, some information on the formations that will be encountered need to be listed, as well as the project specific requirements. You need to bear in mind that when you get in operational mode, you need to have all equipment necessary on board of the rig and not waste time waiting for the equipment to arrive since every minute spent on the rig is extremely expensive. Therefore, selecting the adequate bits for the different phases of the well is a time-consuming and challenging task, in which you need to mitigate all the possible risks that will be encountered and make sure you consider all possible scenarios.

The two types of bits that are mostly used in oil and gas drilling are the roller cone bit, also known as the rock bit, and the polycrystalline diamond (PDC) bits. Their main differences are the following: roller cone bits have three cones that rotate, so three moving parts, its cutters are made of steel or tungsten carbide and are most commonly used for softer formations. On the other hand, the PDC bits have no moving parts, its cutters are made of diamond and are most commonly used for harder formations. A big difference between the two is the price, PDC bits are much more expensive than roller cone bits.

BHA Selection

The BHA configuration jointly with the bit selection is a very complex process, but in this case it also involves other disciplines such as the geologists and petrophysicists. The BHA get a lof of focus because its tools also acquire information from the formations being drilled. Since, it is of interest to the geologists and petrophysicists to understand better the properties of these formations being drilled, there is a bigger pressure when selecting the BHA tools.

The main objective of oil and gas drilling as the name suggests is to find hydrocarbons and be able to produce them. First of all, to find hydrocarbons you need to have tools in your BHA, or afterwards in the wireline logging suite, that enable you to understand the properties of the formation and primarily of the reservoir being drilled. These "magic" tools are incorporated in the BHA and they are namely, LWD/MWD and directional drilling tools, which are in constant communication with surface. These tools provide valuable information for the directional drillers to correct the inclination and azimuth of the well and to follow the planned trajectory. They also enable the geologists and petrophysicists to analyze and observe the characteristics of the rock being drilled and consequently compare with their initial predictions. Obviously, this is an interactive process and since we are not seeing what we are drilling, numerous times surprises are encountered and the plan/strategy needs to be adjusted.

As you can probably grasp by now, the BHA configuration is of utmost importance and, once again, it is a joint effort of different functional teams that have to come to a consensus on what is best for the successful drilling and consequent hydrocarbon discovery of any particular well. One of the great beauties as well as difficulties of oil and gas drilling is that most tasks are multi-disciplinary projects in which different teams get involved, each one with their own expertise and needs, but the keywords here are teamwork and consensus. In the perfect world, the drillers would have their fast and easy well, the geologists would have their slow data focused drilling, the reservoir engineers would have their own data to analyze full of pressure points and extensive good quality reservoir information and the petrophysicists would have a full suite of wireline logs to observe. However, in the real world compromises from every functional area need to be taken and a consensus between all teams needs to be reached. A consensus can only be attained if there is a collaborative approach and if everybody understands the real meaning of teamwork.

Another paramount aspect that cannot be forgotten is communication. Especially in a global and dynamic industry like the oil and gas industry that has individuals from all over the world and most of them from very different backgrounds. Clear and concise communication that everybody can understand, to ensure that nobody gets lost in translation or trapped in cultural barriers. The multicultural and international environment teaches you a lot about communication. Being able to relate and communicate effectively with people from different countries, different religions, different cultures, in fact, different anything is not something that you learn in the university or in any academic manual, it only comes with experience and hard work.

BHA & Drillpipe Configuration

The drillpipe selection is also another big step in the drillstring design, but also the last one. After the bit and the BHA have been selected and determined, making sure they are the best fit for that specific project, we focus on the drillpipe. Once again, in the books and software it is all very easy and we can find the optimum size and dimension for our drillpipe. However, we cannot forget that sometimes what we need is not available in the market and we have to go back to the software and re-do our simulations with the pipe that we have available. This situation is very common in oil and gas drilling, and there are times where these iterations occur several times in the same project.

It is important to mention that the BHA is usually ran in compression during drilling operations and everything above the BHA, which is the drillpipe is ran in tension. Other than the LWD/MWD tools and directional drilling tools, in most cases the BHA is also composed of drill collars and heavy weight drillpipes, which have the sole purpose of making the transition to the drillpipes, but are still part of the BHA.

The main problems encountered in the drillpipe design are always related to the bending moments it endures along the whole drilling process. As it was already mentioned, the BHA takes all of the compression and weight of the drillstring, while the drillpipe is in tension and only suffers the bending stresses. The length of drill collars that are required for a particular drilling situation depends on the weight on bit (WOB) that is required to optimize the rate of penetration of the bit and the buoyant weight per foot of the drill collars to be used.

Since the drillpipe is to remain in tension throughout the drilling process, drill collars have to be added to the bottom of the drillstring. The buoyant weight of these additional drill collars must exceed the buoyant force on the drillpipe, ensuring that when the entire weight of the drill collars is allowed to rest on the bit, then the optimum weight on bit is applied. The WOB varies as the formation below the bit is drilled away, and therefore the length of the drill collars is usually increased by an additional 15%. Hence the length of drill collars is 1.15*L, to allow more flexibility for the driller. Typically, the length of a drillpipe is 30 ft.

Chapter 4: Rig Selection & Well Control

The first three stages of oil and gas drilling are more focused on engineering, but now we go into the money intensive part where procurement and management come into play because it involves a s big percentage of the total budget for the project. The rig selection is definitely a heavy-weight that involves a variety of stakeholders and a complex workflow including drilling engineering specific rig requirements as well as well control expertise, state-of-art technology acquaintance, procurement ranking and sorting and a thorough drilling rig market analysis.

Rig Selection Workflow

The rig selection is always a very hot topic. Why? Because it involves a lot of money. It is, by far, the biggest contract of an oil and gas drilling operation and the most difficult one to close. It is not easy to secure a rig, especially if we are looking for very specific characteristics. The rig selection process is not that easy to explain but I will try to simplify, as much as possible, this complex workflow that involves a lot of stakeholders and a lot of different inputs. I will walk you through it.

Everything starts in the budget. Before we even start, a budget for the project is stipulated. Then from the budget, a list of available rigs is sorted. Afterwards, we start giving ourselves some boundaries to work on, which can be summarized to two variables, rig depth rating and operating parameters, which both need to be determined, and this is when we really start reducing our funnel of adequate rig specifications. The rig depth rating can only be obtained from the casing program, which at this stage is not an easy task to perform since very little information is available. The operating parameters also need to be taken in consideration, which in the case of an offshore project are even more demanding. The operating parameters can be the different water depths (in case of offshore drilling), the weather conditions where the operations will occur in terms of wind and temperature, the location of the operations if in the desert or in the middle of a highly populated city, the environmental sensitivity of the area, rig transportation constraints, as well as other parameters specific to each location.

In order to simplify the process, we can assume that the depth rating for a specific rig can be summarized to, what will be the maximum expected load of the heaviest casing string, which usually is the section of intermediate casing. It is also important to mention that the rig depth rating is usually translated to the hoisting requirements of the rig, both terms are used by the drilling contractors, which is the name given to the rig owners. Once again, taking in consideration an example for easier understanding, if an intermediate 9-5/8" casing string weighs around 413 kips, we would need a rig with a hoisting capacity of 500 kips. Taking in consideration that the intermediate casing was the heaviest load, which always needs to be verified in the casing program but usually it's the case, this would give us enough margin to run all casing strings without any problems.

The next step is to obtain the mud program, which translates into the mud pumps requirements for the rig. This will reduce even more the list of adequate rigs for this specific project. As you can see, it is not an easy workflow and even though there are a lot of rigs out there, this is certainly a very long and complex process.

The solids control equipment is also a requirement for many projects and nowadays even more due to all the environmental and safety requisites that are demanded by the local authorities and, above all, they are the industry best practices that should be followed at all times. Once again, after a thorough evaluation of all the requirements, and bearing in mind this is an iterative process that sometimes goes back to step one, we finally get a list of adequate rigs to analyze our well control needs.

Well Control Needs

Well control is a very sensitive issue that needs to be looked at, in great detail. The pore pressure / fracture gradient predictions, the casing design calculations, the fluids that will be encountered downhole as well as the pressures, all of them have associated uncertainties. All of these uncertainties, need to be taken in consideration when well control comes into play, since this is our last barrier if an uncontrolled hydrocarbon event occurs.

An uncontrolled hydrocarbon event is the greatest fear in oil and gas drilling and one of the main goals of everybody involved in the operations is that something like that, never occurs. The safety first mindset is always present due to past events that have killed many individuals and nobody wants that to happen again. The planning, the safety factors, the attention to detail, is all done with one purpose, perform operations safely from start to finish with zero incidents.

In terms of operations, kick occurrence is monitored constantly by the drilling crew. The one circulation method is the preferred method to be used in a kill operation due to its lower duration: drillers method requires one drill string and two annulus displacements, while the circulation method only requires one complete displacement of the well. A heavy drilling fluid shall be prepared prior to the start of drilling operations and stored in an auxiliary mud tank to reduce time demanded to adjust the fluid density to kill mud weight.

There are many ways to dimension a BOP, in a drilling project, however to be on the conservative side of safety, a good practice is to consider the worst possible scenario. In this case, a worst case scenario would be to ensure that the BOP is able to guarantee that if the well is closed, even in the worst case scenario, for an offshore project, with the well and mudline full of gas after a kick during drilling, there will be no risk of blowout. Assuming, for example, the pressure obtained from this worst case scenario shows that the BOP needs to withstand a pressure of 5,045.15 psi. We then need to go to the market and see what type of BOP's will be needed for this specific well. Normal pressure BOPs (3,000 or 5,000psi) would not be suitable, then high pressure BOP's (>10,000psi) would need to be used instead.

It is important to mention, that this requirement can be a key screening factor to select the adequate drilling rig for a specific project. Every rig comes with its own BOP and sometimes the rigs meet all of the other requirements, depth rating, operating parameters, mud pumps, solid controls equipment, but they do not have the adequate well control equipment. For example, in this scenario that was discussed in this chapter, with a well that has a high pressure reservoir, a rig with a 10,000 or 15,000psi BOP would have to be searched in the market, since rigs with normal pressure BOP's would not be acceptable.

Rig Market

After determining, all of the rig specifications needed to meet the requirements of a specific project, it is time to go to the market, with that short list and invite those rigs to tender. It is very rare to find the perfect match and so compromises will have to be taken and the best available solution is usually selected, even though it most often does not fulfill all of the requirements.

Once again, just like it is done for the service providers and logistics equipment, the evaluation of the tenders need to be performed in commercial terms as well as technical terms. A lot of attention is given to this exercise, since the rig is the most expensive contract of the whole project, so top management gives it a lot of attention and there is considerable pressure involved in the whole process. The rig market is a global market and there are a lot of drilling rigs out there, however a lot of variables need to be taken in consideration. The rig needs to have a time slot to fit our operation, the rig needs to be close if not the rig mobilization to the place where we want to drill our well will take a lot of time and cost a lot of money, sometimes even more than the actual well, which would obviously make the project economically unviable. Therefore, all of these variables need to be taken in consideration, making the whole rig contracting process much more complicated.

The oil price volatility affects tremendously the rig market, making it also very volatile. Obviously, the oil price affects the oil and gas industry and that is why it is a cyclical industry that has its ups and downs. Now, there are a lot of people that think they can predict the future oil price but there are so many variables affecting it that it is impossible to be sure. The oil price is affected by political, environmental, technological, economical, geopolitical, and so many other types of variables that it is not feasible to control all of them and know where all of them are heading. Honestly, the oil price predictors can be considered a sort of psychics because there are so many components of the oil price that they have no clue how they are going to evolve, so they have to guess a bunch of those variables, to be able to actually get to a number for the future oil price.

The rig count, number of rigs operating at a given time period (being it a specific day or month of the year), varies with the oil price and therefore there is a lot of demand for rigs when the oil price is up but when the oil price comes down that demand goes away and the rig count decreases. This is a big adversity for the drilling contractors that need to be flexible and adjust to the volatility of the oil and gas industry. It is also a challenge for the operators, when the oil price is high, since there is a lot of demand for rigs, the drilling contractors increase their prices because they will always manage to find work. Consequently, there is always a contest between operators and drilling contractors to get the best deal when signing these rig contracts and the volatility of the oil price just makes it more challenging, as well as, more entertaining.

Chapter 5: Logistics Strategy

Logistics is that slice of the pie that is disregarded most of the times but it involves a lot of hard work that usually passes by unnoticed, if everything goes smoothly. However, it has everything to go wrong due to its complexity, starting with the huge number of service providers that need to be contacted and all of the equipment that needs to be sourced. Contracts need to be put in place, evaluated and negotiated to ensure we are getting the service we desire and that we are paying a fair price for it. In this case, as it is done for the rig contract, the technical teams and procurement teams need to work hand in hand to ensure requirements from both teams are met, consequently being the best fit for purpose solution for their company. The logistics involves, looking at the big picture, all of the transportation, storage and offloading for operations of all the consumable and tangible equipment to and from the rig, as well as, the personnel to and from the rig, and not forgetting the mobilization and demobilization of the rig.

The mobilization and demobilization is a very important part because it is usually where accidents happen, people tend to relax because they are not in operational mode, and when people relax that is when incidents tend to occur. The same thing happens at the end of each shift, when the mind starts drifting to what you will be doing after the shift and the focus on what you are actually doing is not there anymore. The mobilization and demobilization involves a lot of hard work and since it is only done every once in a while, sometimes things go wrong because people are not used to that routine. Once again, the attention to the detail and the concentration need to be at their peaks to ensure all operations are performed in a safe manner with zero incidents.

Port

All the heavy and state-of-art equipment needs to be transported, stored and ready to be delivered to the rig at a specific time when operations dictate so. The rig also needs diesel and drilling fluid on a daily basis. The operations occur twenty-four hours per day, seven days per week, so consequently all off the logistics around it needs to be ready to work and to deliver at any time of the day or night. Taking in consideration, that all of this equipment is quite heavy and that the traffic on the quayside and around the port will be intense, the port needs to be prepared to sustain heavy equipment. Unfortunately, in remote locations where a lot of oil and gas drilling occurs, the infrastructure is not ready for any of this. Therefore, special attention needs to be taken in frontier areas and site surveys/investigations need to be performed to ensure all best practices are being followed. In previous occasions, ports have been found to be full of "hidden" pot holes or have been built in extremely bad conditions, so their foundations were terrible and not able to sustain any kind of heavy equipment. Mitigation measures had to be put in place, however if this special attention was not taken in the first place, unnecessary incidents could have occurred. Once again, all of these precautionary measures have the same goal, that all operations are performed in safe conditions and everybody goes back home to their family with their ten fingers, two arms and two legs.

The port is, in most cases, a big issue because a contract needs to be put in place with the local authorities of the port, which involves a lease of a certain area to enable operations to run smoothly. This certain area is usually a big portion of the port, especially in frontier countries. All of this process takes a long time, identifying the appropriate area to perform all necessary operations, speaking with local authorities to understand if it is feasible to use that certain area, surveying the whole area that will be used, ensuring the local communities will understand the usage of that area and won't be a burden, coming to terms with local authorities and local communities in terms of contractual decisions, educating local communities of what we are going to be doing. After all this is approved, the next step is to find local content to work at the port and train them appropriately, as well as, ensure everybody is working with appropriate Personal Protective Equipment (PPE) at all times. Any operation performed at the port needs to be done in a safely manner making sure all operations in the port run smoothly.

Transportation

The transportation can be divided in two parts, the transportation of human resources and transportation of equipment.

First, we have all of the people travelling in and outside of the country, where the operations are taking place. The operational teams also need to travel to and from the rig, and if it is an offshore project it will all be done by helicopter.

Moreover, if it is an offshore project, it also has support vessels, at least three around each rig that are constantly going from the port to the rig and back, bringing and taking equipment, drilling fluids, fuel and in some cases switching crews. The process of loading and offloading drilling fluids, fuel and equipment on a daily basis at the port and also at rig site involves a great deal of hard work and expertise to be set up and to run smoothly. A lot of different companies are involved in the whole process, each one of them providing their own specialty service. The process needs to be managed by company representatives that coordinate the whole process and ensure everybody is being taken care of and safety rules are being followed at all times. The transportation of goods in and out of a country also involves a lot of back office work in terms of customs clearance, taxation framework, dealing with local authorities during this whole process, as well as, ensuring all goods coming in and out of the country are treated following all safety procedures.

Operations Support Base

The operations support base is where all the engineering and planning occurs in support of the drilling and logistics operations. It is usually an office that is set up on a city close to the operations and to the port supporting the offshore operations. The people that stay in that office have a daily morning meeting with the people on the rig. Usually, you have at the operations support base, representatives of the oil and gas company, usually known as the operator, of the drilling contractor, which is responsible for the crew that operates the rig, as well as the main focal points of the service providers such as Schlumberger, Halliburton or Weatherford. These service providers, as the name suggests are specialized in specific drilling operations services such as directional drilling, MWD/LWD, cementing, wireline logging, mudlogging, drilling fluids and casing running equipment.

Chapter 6: Plug & Abandonment

The plug and abandonment of an oil and gas well is a procedure that has existed since the early days of the industry. However, nowadays the number of wells being plugged and abandoned has increased exponentially. Consequently, the corresponding expertise on this subject has increased, as well as, the attention from legislative authorities to ensure the oil and gas industry is doing it correctly. The association of this operation with the environment is evident, since it is the last operation performed on an oil and gas well and, therefore it has to be performed correctly, to ensure its surrounding environment is not damaged in any possible way.

Theoretically, the final abandonment of each well involves pumping cement down the well to provide the isolation with cement plugs in the various hydrocarbon-bearing zones as well as aquifers. In some cases, mechanical plugs are placed below these cement plugs to serve as a structural base for these plugs and to avoid any slumping of the cement before it sets. Cement plugs are also placed at liner tops and near the surface, or in case of offshore wells just below the seabed level. Another important section of an adequate plug and abandonment process, and that sometimes is forgotten, is the surface equipment, which should in all cases be recovered respecting all of the regulations and discarded appropriately, minimizing its impact on the environment.

In summary, anywhere in the world, regulatory bodies have, to varying degrees, defined procedures and responsibilities for a plug and abandonment of any specific well. Despite the differences in legislations throughout the world, the objective of all abandonment operations can be summarized in the following four points:

- ✓ Isolate and avoid the intrusion of fluids in any water pockets.
- ✓ Isolate any potential permeable or commercially viable section.
- ✓ Avoid any potential leaks into or out of the well.
- ✓ Cut casing or any pipe to a stipulated depth below seabed (for offshore wells) or below ground level (for surface wells) and take away all equipment at seabed or surface.

Best Practice Guidelines

The industry best practices are always an arguable topic because it is a subjective matter that depends on each individual perspective. It is clear that after the well has been drilled, it has to be plugged to ensure that there isn't any unwanted migration of hydrocarbons to surface. It also has to be abandoned in a safe manner to leave the environment around it unaffected by its presence. However, each person has their own view on how this process should be done and that is why, it is always a very sensitive topic. A very good reference to keep in mind are the U.K. Oil & Gas Guidelines for abandoning a well.

These guidelines state that any zone that is permeable needs to be isolated from the surface (or seabed) and from each other by at least one Permanent Barrier. Another main requirement is to have two Permanent Barriers between surface (or seabed) and a permeable zone that has hydrocarbons or is water bearing but over-pressured. Note that, when these guidelines refer to a "Permanent Barrier" they mean a tested plug of good and sound cement of at least 100 ft. The U.K. Oil & Gas Guidelines also state that offshore wells must be abandoned leaving no casing until 10 ft below the seabed level, to avoid disturbing the sub-aquatic fauna. They also refer that all subsea equipment should be removed when possible and that none of this subsea equipment should represent any danger to other marine users.

Another good set of guidelines are the NORSOK directories to abandon a well. These are used by Norway for all of their oil and gas activity, and they are also very complete and simple to understand. Both of these guidelines are not free of charge, but it is possible to acquire them and get acquainted with all of its details.

Abandonment Procedures

The abandonment program needs to be developed, pursuing the abovementioned industry best practices and focusing on the best technical performance, as well as, cost effectiveness. In some cases, we are not only abandoning the well but we also have to decommission the rig and ensure it is removed in a safe manner and leaving the smallest amount of environmental footprint. Therefore, following the best practice guidelines, all the structures and facilities abandonment, rigged in water depths lesser than 250 ft and 4,000 ton of weight in air excluding the deck and superstructure, should be completely removed. In cases, where foundation piles were used, 10 ft below the seabed, will be left in place.

In case we have production wells, they would all be previously and effectively killed to prevent the potential oil leakage to surface. Considering a possible standard program for clarification purposes, it would consist on perforating and cementing casing and production tubing in place, establishing a set of 4 cement plugs (permanent barriers) and making sure those cement plugs were evaluated using the CBL method, which was previously explained, to ensure the quality of the cement placement. Then, the casing strings would be cut 10 ft below the seabed. This could take a minimum operational time of 4 days and it would reduce considerably the potential of risk hazard, and obviously eliminate the handling and treatment of radioactive and scaled tubing. Another important HSE advantage, that should always be performed is to ensure the removal of the production wellhead and corresponding safety valves, is only done when the wells are completely deactivated.

Abandonment Options

There are two main options for abandoning wells: abandonment in place and full abandonment. What I have been discussing is the full abandonment, however in some cases an abandonment in place is performed because the operator will return to the well in a short period of time. There is always a decision-making process involved in the abandonment process, to decide what are the permeable zones that need to be isolated, what is the casing that will be cut and removed, how many feet below the seabed will it be cut, how many cement plugs are we going to perform, how many feet of cement plug are we going to pump for each cement plug, and how are we going to test those cement plugs. All of these decisions need to be discussed and made, but the U.K. Oil & Gas Guidelines exist, precisely for that reason, to aid in all of this decision-making process.

Chapter 7: Environmental Impact Assessment

The environmental impact assessment (EIA) is an important tool designed to identify and predict the impact of a project on the bio-geophysical environment, human health and socio economic aspects. This process starts from the front-end engineering design stage until the final decommissioning. It is used to interpret, analyze and communicate information about the potential repercussions of the project. As well as, to provide alternatives, possible solutions to manage and mitigate negative impacts and to maximize the positive ones. Environmental disturbances are in general more expensive to correct after their occurrence than before. Thus environmental issues must be addressed as soon as possible during project planning. Likewise, economic, financial or technical studies, the EIA is an integral part of the project. There are two main assessment aspects in an EIA, first is the significance of the event that is being evaluated. A significance matrix has been used as an industry standard to classify these events in which you classify each impact in terms of its magnitude and you also classify in terms of importance/sensitivity/vulnerability this impact has to the receptor or to its source. This EIA significance matrix is a powerful tool to prioritize each possible event.

After determining the significance of each possible event, the corresponding mitigation measures need to be determined. One of the key objectives of an EIA is to identify and define socially and environmentally acceptable, technically feasible and cost-effective mitigation measures. Mitigation measures are developed to avoid, reduce, remedy or compensate for the significant negative impacts identified during the EIA process, and to create or enhance positive impacts such as environmental and social benefits. In this context, the term, mitigation measures, includes operational controls as well as management actions. It involves the whole spectrum of the organizational chart of a project from the person at the utmost top to all of the workers involved in the project.

Mitigation measures are often established through industry standards and may include the following:

- ✓ Changes to the design of the project during the design process (for example, changing the development approach or selection of more energy efficient power generating equipment).
- ✓ Engineering controls and other physical measures applied (for example, use of effluent treatment equipment or spill prevention technology).
- ✓ Operational plans and procedures (for example, notification to other marine users in case it is an offshore operation, navigation safety plans or waste management plans).

Uncertainty plays a big role in this whole process. Even with a final project description and an unchanging environment, predictions of impacts and their effects on resources and receptors can be uncertain. Predictions can be made using varying means, ranging from qualitative assessment and expert judgement through to quantitative techniques (for example, discharge modelling). The accuracy of predictions depends on the methods used and the quality of the input data for the project and the environment. This input data is usually provided by the operator to the specialist company that was hired by the company to simulate, execute and conclude the EIA process under the operator's supervision.

Where uncertainty affects the assessment of impacts, a conservative (for example, reasonable worst case scenario) approach to assessing the likely residual impacts is adopted and mitigation measures developed accordingly. To verify predictions and to address areas of uncertainty, monitoring plans are normally proposed. The main goal of an EIA is to assure that all stakeholders are aware of the risks that are being taken and that all possible mitigating measures are being undertaken.

HSE Bridging Documents

Nowadays and fortunately, the Health, Safety and Environment (HSE) is one of, if not, the main focus when drilling for oil and gas. It is of paramount importance that everybody has the HSE mindset when involved in oil and gas drilling. Therefore, an EIA is done for every well that is ever drilled and this is a big evolution from how things were done 15-20 years ago. As already mentioned, an EIA involves a complex survey with various entities to ensure that all mitigating actions are being performed to avoid any environmental accidents. It is also important to acknowledge that all catastrophic scenarios are considered from an environmental point of view.

An important part of this complex project is to ensure that bridges are built between the safety management systems of the different stakeholders. These are called the HSE bridging documents, which bring together the HSE focal points of the operator and of the drilling contractor to ensure that they agree on the terms of the HSE bridging document and decide which entity is responsible and liable for each specific eventuality, and everything needs to be documented on paper. Once again, it is a long and complex process that involves a lot of negotiations, the legal parties also get involved to ensure their entities are abiding by the law, as well as, the technical staff to ensure all industry best practices are being followed. It is important that everything is written down and signed before the operations start, since it needs to be defined who are the responsible persons in case an environmental incident occurs.

A thorough and detailed risk assessment involving all focal points from all functional areas is also performed to ensure all stakeholders are aware of the risks and to perform a sort of a brainstorm to bring all of the different ideas to the table. Having together, focal points from drilling, geology, reservoir, logistics, finance, legal, tax, risk management, asset management, projects, flow assurance & process, petrophysicists and geophysics, is not an easy mission but it brings a lot of different perspectives into the discussion.

Environmental Risk Assessment

The results of the identification and the ranking of environmental and socioeconomic impacts and risks are usually summarized in a table, usually called risk assessment matrix, where the risk is classified in terms of likelihood to happen as well as the level of its impact and the effect of the mitigation measures that should reduce both the likelihood of happening as well as its impact. The risks identified could arise directly or indirectly from routine and emergency situations during the lifetime of a specific project, which are usually split in four stages: drilling, installation, production and decommissioning, we have been speaking only about oil and gas drilling, however, in this section, I decided to include all of the project stages to give you a better sense of the big picture when we are talking about an oil and gas project. Especially, for the EIA, the project needs to be looked at like a whole, end-to-end, if not this exercise does not make sense.

The methodology that is normally applied is the following, a specialized company is hired by the operator to analyze all of the environmentally sensitive situations that could occur with the specific operation. This specialized company demands a huge amount of questions to the operator in terms of diesel consumption, helicopter flights, drilling fluid consumption, food consumption from the people on board the rig, sound emitted and various other details that are really specific to the project. From there, they introduce all of this project specific data into their specialized computer models to calculate the impact the operation may have on the environment.

Afterwards, an assessment is performed to see what would be the outcome of each mitigation measure and how they would reduce the impact of certain events, always with the same goal in mind of reducing all risks to as low as reasonably possible (ALARP).

Considering a hypothetical project, after taking the effects of the planned mitigation measures into account, no high environmental risks should be identified during the assessment. The risk assessment usually identifies activities associated with the project with a medium risk associated to them. Three impacts, which are usually addressed with special attention are:

1. Long-term physical presence of the facilities;
2. Fishing activity and local tourism;
3. Accidental hydrocarbon release.

The long-term physical presence of the facilities, usually in projects that involve production, need to be evaluated to ensure the bio diversity of that zone is being affected as low as possible.

Fishing activity for offshore projects as well as local tourism need to be taken in consideration to ensure the local population is affected as little as possible. In fact, many offshore fields are in the border of fishing areas and most of their areas of influence contain artisanal fishing activity. From the Marine Regulations, a security zone of 500 meters around the installations will be delimitated to assure operational safety and avoid any type of accidents. In these security zones around the offshore facilities, fishing should be forbidden. Obviously, this is a safety measure that needs to be followed, but unfortunately, it can potentially create conflicts. Therefore, specific programs for communication with the local fishing community are put in place to create awareness of the activities being performed and to explain why fishing will be forbidden in those security zones.

Moreover, pipelines and subsea structures could also have the potential to limit access for fishing and impede fishing gear. In any case, the environmental entity responsible for the EIA always notifies the local authorities, which will issue notices to mariners to advise fishing and shipping traffic of the potential hazards to navigation that are associated with an offshore project in their fishing areas. Mariners will be advised of specific periods and locations in which vessel operations should be avoided.

Although the probability of occurrence is very low, the emergency events are also considered as being of significance to the environment. Indeed, they can result in a complex and dynamic pattern of pollution distribution and impact in the environment. The environmental impact of a spill depends on numerous factors including:

a) Location and time of the spill.
b) Spill volume.
c) Hydrocarbon or chemical properties.
d) Prevailing weather conditions.
e) Environmental sensitivities.
f) Efficiency of the contingency plans.

A model of oil dispersion evaluates with precision the risks for conservation units or sensible coastal ecosystems. If necessary, the main forces in the hydrodynamic model will be monitored in the area of influence of the activity. In case of emergency, the procedures, equipment and time response will be planned to avoid contamination and sensible areas.

In any oil project, all the installations and facilities have Emergency Response Plans (ERP) and Oil Spill Contingency Plans (OSCP). These plans will be prepared based on detailed definition of the areas of influence, risk assessment tools and the EIA. Based on this study, the ERP will be prepared using each accidental hypothesis raised such as fire or explosion, chemical leak, oil spillage and watercraft collision. Even though, technology and processes have developed throughout the years to avoid more and more these situations, there is always a possibility of happening. Therefore, OSCP's and ERP's are put into paper and simulations, as well as, drills of these plans are performed prior to the start of operations.

Oil Spill Contingency Plan & Emergency Response Plan

Every time a well is drilled, an OSCP needs to be put in place to ensure that if the worst happens, there are mechanisms that can be activated to reduce the effects and consequences to a minimum. The oil and gas industry has quite a few companies that specialize in these activities, such as, OSRL – Oil Spill Response Limited, WCS – Well Capping Systems or Wild Well Control.

These companies have specialists that have worked on extreme situations, involving environmental spills, during their whole lives and they are, without a doubt, the best at it. Hopefully, you will never need to request their services but a whole preparation and planning needs to be performed either way. Various scenarios need to be considered from catastrophic events to small problematic situations. Simulations in industry specific software are performed to understand the magnitude of the effects and what mitigating actions can be performed to reduce the consequences.

An ERP is also put in place to ensure everybody is informed and everybody in charge or responsible for a specific function knows what to do, who to call and what mechanisms to activate. Since the operations are always running no matter what, twenty-four hours a day, seven days per week, there is always someone on call and there are always people in charge to activate the ERP, if needed. This plan involves people from the operator side, from the drilling contractor side, from the local hospital side, from the local authority side, from the service providers side and from the transportation side. All of these stakeholders are located at the rig, at the nearest local city, at the airport, at the operations support base and also at the several company headquarters.

As previously mentioned, situations such as kidnapping, fire and explosions, natural disasters, and any other catastrophic events need to be considered and contemplated in the ERP. All emergency contacts need to be listed as well as all of the tasks outlined for each intervenient. All participants need to have a backup and be aware that when they are on call, they need to be ready to pick up the phone at any time of the day or night. Once again, it is a very detailed document that everybody hopes and works every day for it not to be used, however it is extremely important that it is put in place and that the utmost attention is given to its make before any operations begin. In summary, the safety mindset has to be present at all times, since at the end of the day it is the most important aspect of oil and gas drilling. It is all that really matters. Never forget – safety first.

About the Expert:

I am a Mechanical & Petroleum Engineer (dual masters) with five years of work experience, always representing the same O&G company and doing so in four different countries: Portugal, Namibia, Morocco and Brazil. I am fluent in four languages: Portuguese, English, Spanish and French.

The international experience during my childhood where I lived in Italy, Brazil and Argentina (other than Portugal) aided me in thinking out of the box. The two exchange programs I participated in Sweden and Argentina reinforced this situation and helped me understand that an international environment incorporated with strong teamwork is definitely the key to success.

Travelling is one of my big passions, I have travelled a lot during my childhood and I have travelled a lot for my job. Luckily, I also have a competition with my wife, which is to visit all the countries in the world, we are passed the sixty countries and we want to reach the seventy countries mark ASAP!

HowExpert publishes quick 'how to' guides on all topics from A to Z by everyday experts. Visit HowExpert.com to learn more.

Recommended Resources

- HowExpert.com – Quick 'How To' Guides on All Topics from A to Z by Everyday Experts.
- HowExpert.com/free – Free HowExpert Email Newsletter.
- HowExpert.com/books – HowExpert Books
- HowExpert.com/courses – HowExpert Courses
- HowExpert.com/clothing – HowExpert Clothing
- HowExpert.com/membership – HowExpert Membership Site
- HowExpert.com/affiliates – HowExpert Affiliate Program
- HowExpert.com/writers – Write About Your #1 Passion/Knowledge/Expertise & Become a HowExpert Author.
- HowExpert.com/resources – Additional HowExpert Recommended Resources
- YouTube.com/HowExpert – Subscribe to HowExpert YouTube.
- Instagram.com/HowExpert – Follow HowExpert on Instagram.
- Facebook.com/HowExpert – Follow HowExpert on Facebook.

Printed in Poland
by Amazon Fulfillment
Poland Sp. z o.o., Wrocław